Henry McElderry Knower

The Embryology of a Termite, Eutermes (rippertii?)

Including a contribution to the discussion as to the primitive type of development, and the origin of embryonic membrands [amnion] and of the mesoderm in the Insecta

Henry McElderry Knower

The Embryology of a Termite, Eutermes (rippertii?)
Including a contribution to the discussion as to the primitive type of development, and the origin of embryonic membrands [amnion] and of the mesoderm in the Insecta

ISBN/EAN: 9783337325046

Printed in Europe, USA, Canada, Australia, Japan

Cover: Foto ©berggeist007 / pixelio.de

More available books at **www.hansebooks.com**

The Embryology of a Termite —
(*Eutermes (Rhinotermes?)*)

A Thesis —
Presented to the
Board of University Studies
in the
Johns Hopkins University —

For the degree of
Doctor of Philosophy —

By
Henry McElderry Knower. A.B.

Baltimore —
May. 1896 —

The Embryology of a Termite
with one plate (?)

Though the Termites must be ranked among those insects which have best preserved ancestral traits, their development has not been studied up to the present time — Indeed, it is only within the last few years that we have been enabled to judge of the details of development of any of the primitive forms of insects — The technical difficulties which the investigator of these forms must meet, are very great, much time being required to obtain any results, especially in dealing with early stages — As Wheeler (20) has suggested, no doubt this has deterred many from working on such material — I have devoted considerable attention to

the early stages of the development of the termite in the hope that such a study might throw light on some vexed questions of insect embryology. The later stages can not be treated in detail in the present paper, but will be described at another time —

* The eggs were obtained in Jamaica at different times, by myself, and by the kindness of Dr Hough, and Mr Sigerfoos. I am also much indebted to Mr Taylor of Jamaica for assistance in obtaining material, and to Dr McCatten for many suggestions as to technique. It is a great pleasure to express my appreciation of the constant kindness, and valuable criticism and advice, of Professor W. K. Brooks, during my

x This species from Jamaica has been described as Eu-Termes Rippertii — There is some doubt in my mind as to the species, hence the mark of question in the title —

work at the Johns Hopkins University — I must also remember here, with sad regret for his loss, the constant and stimulating friendship of the late Colonel Marshall MacDonald, by whose invitation I was enabled to spend some time at the United States Fish Commission laboratory at Woods Holl, Mass.

There are no special chambers for the queen or for nurseries in the nests of this species, but the inner parts of the nest differ somewhat from the more peripheral portions — The outer series of passages of the Termitarium having somewhat thicker walls than the inner ones —
The queen is generally found in the interior of the nest, near its base, surrounded by numerous workers and larvae, and

not far from the heaps of eggs. These eggs are collected by the workers as they are laid, and piled up in the passages near the queen, without any apparent system. The older larvae also take part in tending the eggs. It is an easy matter to collect the eggs, since they are quite moist and adhere together in masses, in which all stages are to be found, from unsegmented ova to larvae leaving the egg membrane.

My material was fixed with hot water, with cold picro-sulphuric acid, and with hot picro-sulphuric; the latter method giving the most satisfactory histological results.

The egg is elongated, larger at one end and markedly

convex on one side — This shape makes it easy to determine the planes of symmetry from the start, since the enlarged end is the micropyler or posterior pole while the convex side is the ventral surface — In the course of its development, the embryo changes its position by a remarkable process of revolution, like that described for the Libellulids and certain Orthoptera. This must be kept in mind in speaking of the anterior and posterior poles and of the dorsal and ventral surfaces — As used above, the micropylar end of the egg is the definitive posterior pole, and the convex side is the final ventral surface —

Dr Hagen (¹) described the micropyles of the eggs of this species as follows: "The micropyles of Termes' eggs have not

* Numbers in parenthesis after a name refer to references in back of book —

before been known — near the upper pole of the egg, on each side, there are four to six flat impressions, viewed in profile they are similar to a soup dish, in the middle of this shallow funnel is a tube of smaller diameter going through the yolk-membrane in the direction of the egg pole." This description must be modified, since I have found that the funnels are more numerous and are grouped on the ventral surface — The chorion is the only membrane penetrated by these funnels — The micropyles are arranged in a semicircle, on the convex face of the egg, near the posterior end — This semicircle is composed of from 12 to 18 funnels, which vary considerably in their arrangement — They may be strung out into a single line,

curving over the ventral surface, just above the place where the embryonic disc first appears, and extending up on either side towards the dorsal surface — After the funnels at the two ends of the semicircle are crowded together while the ventral ones form a single row — Fig. (1). In some eggs, on the other hand, the median ventral funnels are crowded, while the lateral openings are strung out into a line — In surface preparations, the mouths of the micropyles may be seen to be slightly elongated — Dr Hagen ( ) thought he could "see distinctly a brush of filiform spermatozoa inside the micropyle in some eggs," and that "in one case he had succeeded in bringing them out by pressure", but it is extremely doubtful that what he

saw one spermatozoa — Filiform bunches often appear beneath the chorion, but though I have looked most carefully through a number of stained miciegmented ova, I have not succeeded in finding any spermatozoa — The objects Dr Hagin mentions were probably folds of the delicate vitelline membrane, possibly hyphae of a parasitic fungus — In surface views of alcoholic specimens, the folds of the vitelline membrane and protoplasmic threads attached to the chorion, at times, seem to be bunches of filiform bodies — When such specimens are crushed, the membrane wrinkles-up, and looks like bunches of filaments — Eggs of my material from one nest had been attacked by a fungus to such an extent, that

the hyphae entering the micropyle had ramified in the yolk, and used it up, destroying the embryo; but Dr Hagen could hardly have mistaken such large objects for spermatozoa —

Yolk —

My observations on the yolk were made chiefly on preserved materials — It is composed of a mass of flattened polygonal bodies, which are vesicles containing a homogeneous coagulable fluid — These bodies stain deeply in Haematoxylin and Carmine and vary considerably in size and shape — The other constituent of the yolk-mass is formed of oily globules, most of which generally collect into one or two quite large drops — When alcoholic specimens

are crushed, the oil globules run together and flow out, leaving the yolk-bodies for the most part intact — in such preparations the character of the vesicles can easily be made out — They are found to vary in structure, part of them being entirely filled with the homogeneous coagulated substance, which is quite elastic under pressure — many of the bodies are however different from this — They also contain the homogeneous stainable substance, but in addition little oily drops, of which there may be only a few or a great number — The greater the collection of these drops the less is the homogeneous material; so that it is often reduced to a porous substratum for the drops in the centre, and a thin layer around

the periphery of the vesicle — It is possible to crush such vesicles, so that the oil drops escape through the membrane surrounding the vesicle — Examined in clove oil, or in sections, the yolk has quite a different appearance — In these specimens the oily matter is all dissolved out, even from the vesicles — The treatment with oil often causes the the yolk-bodies to fuse into large masses — In sections this is often very marked, and large spaces are formed through the mass by the solution of the oily substances — Sometimes however a section of a single yolk-body is met with, riddled with holes left by the solution of the oil drops which were imbedded in the substance of the vesicle — Figs. (23 & 28) — Agreeing with

Wheeler's views of the yolk of Blatta, (3) I should regard the "granular" looking yolk-bodies, containing oil drops as derived from the homogeneous variety — The homogeneous coagulable fluid of these bodies becomes gradually transformed into fatty substances, and these at first being formed in isolated drops, finally fill the whole vesicle — In this way the oil globules and free oily fluid of the yolk would appear to arise from the albuminous (?) bodies to furnish the growing embryo easily assimilable nutriment — When yolk-bodies in which drops are collected are preserved for sectioning, by the usual methods, chemical changes take place, which result in homogeneous fixed masses of stainable (albuminous?) substance, and

in the solution and removal of the fatty
matter —

There is apparently no definite arrangement
of the different yolk elements —

There is no peripheral layer of cytoplasm
distinguishable, before the formation of a
blastoderm —

I have found no segmentation of the yolk
during the early stages of development —

Segmentation and formation of first rudiment
of the embryo —

The first sharply marked rudiment of
the embryo is, as is the case with certain
of the Orthoptera (Stenobothrus, Stagmomantis,
Gryllus, and Oecanthus), a relatively
small disc of closely crowded cells at
one pole of the egg —   Since the

history of the origin of the embryonic discs has not been worked out, I have studied, with special care, the segmentation and the changes in the blastoderm cells on the surface of the yolk, which result in the formation of the embryonic rudiment.

The questions which I have endeavored to answer are: Is the disc formed immediately during the segmentation by cells wandering directly to their places, on the surface, at the point where the disc is to appear? Or is a blastoderm first formed as a result of segmentation, and then the disc formed from its cells? Again, if the latter is the case, is the embryo the result of a simple multiplication of the cells of a restricted area of the blastoderm; or is there some other factor present in the formation of

the disc? The answer to these questions is of ~~considerable~~ importance; hence I have devoted considerable time to it, believing that it furnishes the only sure basis for an understanding of the origin of the 'medullary-loops' and of the amnion — I shall try to make clear that a careful study of the early changes which take place in the blastoderm is a necessity to a consideration of these later formations —

The position of the polar-bodies marks the dorsal pole of the shorter axis of the egg — I have only seen five or six eggs which showed, indistinctly, the formation of the polar-bodies from the nucleus —
When the segmentation nucleus has returned to the centre of the egg, after the extrusion of these bodies

they are seen as two little rod-like masses of chromatin surrounded by a small quantity of protoplasm, lying at about the centre of the flattened dorsal surface — Figs. (2) and (3). Later the chromatin breaks into fragments, but the little collection remains visible for a number of divisions —

The segmentation nucleus, after moving back from the dorsal surface, lies, just previous to the first division, in the centre of the yolk, at the intersection of the shorter and longer axes of the egg — Fig. (2). The first spindle lies at right angles to the shorter axis, so that one of the cells arising from the first division wanders toward the enlarged posterior pole, where the embryo will first appear —

The other cell remains near the position which the mother nucleus held — Fig.(3) — At the start then, there is a decided proliferation toward the future embryonic area. This is brought out better in the following stage, which exhibits two cells in the enlarged end, one in the shorter axis and one in the small end; that is, there are three nuclei nearer the posterior than the anterior pole — Fig.(4) — The cleavage becomes irregular with the eighth cell stage, one or more nuclei dividing before the time for a typical series of divisions — For several divisions there is a slight preponderance of cells in the larger end of the egg — For instance, one egg has four nuclei in this end, one in the shorter axis and three in the anterior end while

another we find no the posterior and four in the opposite end — Fig (5) — Generally, during the early stages of cleavage, there are three or four more cells in the larger than in the smaller end of the egg — After five or six divisions, however, though the number of cells in the posterior is slightly greater than that in the smaller end, owing to the more extensive surface, they have taken up positions at about equal distances apart — It will be noted that I speak of the nuclei with the little masses of protoplasm around them, in these stages, as cells — As far as can be determined there is no protoplasmic continuity between these cells — Later, when the embryonic disc begins to appear continuity is established between its cells, but

over them there is no proof that the blastoderm cells elsewhere on the yolk are connected with one another or with the yolk-cells. A view of the ventral surface of an egg at this stage shows very well the equal distribution of the nuclei on that side, and the same is found to be true of all the cells in the egg — Figs. 6 1, o (A) and (B) — Most of the cells have now reached the surface, there being only a few in the yolk, which are at about equal distances apart.

In properly prepared material, the changes that follow and lead to the appearance of the embryonic disc can be most distinctly traced in entire, transparent eggs studied in clove oil, cedar oil, or balsam. The following description refers chiefly to specimens studied in

this way, andstructions through certain stages —
I have already stated that the various
stages are mixed together indiscriminately
where collected — Hence, the series illustrating
the growth of the disc, has been picked
out of a great mass of materials — there
can be little doubt however that I have
figured a typical series, for the figures
are based on a great many specimens —
Since the first rudiment of the embryo
is formed from the surface cells alone,
the few yolk-cells may be neglected
in the description — The cells on the
surface of eggs with more nuclei than
those just described are slightly more
numerous in the posterior end — In many
specimens, it is evident, that the cells of
this end are dividing more rapidly than
are those in the other end — The nuclei

are grouped in twos on the surface of
each specimen, and are frequently
to be found, at all points in the
surface, in the act of dividing; but
it is in the enlarged posterior end that
the greater number of dividing nuclei
and pairs of just separated cells appearing. (figs. (8). and (9) —

Figures (10, and (10ᵃ) represent the ventral and
dorsal surfaces of a somewhat older egg
than that just described — The number of
cells is now greatly increased, especially in
the posterior end — The nuclei are still
evidently grouped in twos, and some are
seen in the act of dividing — The anterior
half of the egg contains very few nuclei — It is so bare of them, in fact,
that I was for a long time afraid that
their absence was only apparent, and

due to improper handling of the material — This can not be, however, since special care was taken to prick that end of the egg, to insure penetration of the staining fluids; and since the nuclei, which do appear in both ends of the egg, stains very satisfactorily — There is no reason to believe that all of the nuclei are not stained in these stages — It is important to note, in figs. (10) and (15a), how far the numerous nuclei of the posterior pole extend anteriorly — In both the ventral and dorsal surfaces a considerable area is covered by nuclei, at no great distance apart — A cap of nuclei is thus formed over the posterior end of the egg, which stretches forward almost to the shorter axis — The nuclei are

not grouped around any one center.
The eggs shown in figs. (18), (18ᵃ) and (18ᵇ), have more cells, and so therefore somewhat older than that exhibited in the preceding figures, but there is also a difference in their arrangement. In fig. (18) of the ventral surface. the cells of the posterior end are seen to be closer together than in fig. (10), though their numbers have not increased greatly. The anterior edge of this area is more distinctly shown than in the previous stage, and is nearer the anterior end than the shorter axis of the egg. The dorsal surface fig. (18ᵇ) is (comparatively) poorer of cells than in the last stage. At its extreme posterior limit the cells are crowded closely into two or three concentric rows, rather sharply separated from the rest

of the cells of the surface — This can not be shown better in a drawing, though it is much more distinct in rolling the egg — These rows of nuclei mark the posterior edge of the area of closely placed cells, so prominent a feature of the ventral surface — It will be seen from these facts that this area, while quite extensive, is decidedly less so than in the preceding stage — it is most marked on the ventral surface, but extends around on either side and over the posterior end toward the dorsal surface — Fig. (18ᵗ) shows this from the side — Fig. (?) gives a ventral view of an egg at about this stage, in which the nucleated area terminates abruptly at the pole of the egg, and at the sides is seen to thin out rapidly,

as in Kolack figure – fig (18ᵃ) –

In older specimens the nuclei are still more closely crowded, and form a more and more distinct area, which becomes gradually more restricted in extent – Figures (15–18) illustrate this point, and it is also evident in these figures, that the nuclei of this region are of the same size as those elsewhere on the surface – Figures (18) and (19ᵃ) of the same egg, and figs. (15), (16) and (18) exhibit a stage in which the posterior and lateral borders of the nucleated area have become quite sharply defined – This region is now very distinctly outlined, and may henceforth be spoken of as the embryonic area or germ-disc; though it is shortly this until it has reached the stage represented in fig (28) or (27) –

In fig. 27, drawn from an egg tipped up somewhat, the embryonic disc is seen to be a well defined area of nuclei which are quite closely crowded — Careful study shows that they are about the size of those in the surrounding Blastoderm — Anteriorly the disc is not so sharply delimitated, as on the rest of its circumference, and this is true until a late period in the formation of the disc, as will be seen on referring to the figures of later stages —

The facts here reviewed appear to me to show, that the embryonic disc is not formed directly in the segmentation, by cells wandering toward a predetermined point — The evidence seems to indicate also that the disc is not the result of

only active cell multiplication in a restricted area of the blastoderm, but that there is a concentration of cells of the blastoderm toward the place where the first rudiment of the embryo appears — This is shown in the whole series of stages figured, and is brought out vividly by comparison of figs (12), (13), (15) and (25) — The diminution in the area covered by nuclei more closely crowded than elsewhere on the surface is very marked from fig. (14) to fig. (25). In fig. (14) it spreads over the whole ventral surface of the posterior end of the egg, while in fig. (25) it hardly covers one half of this area — The best proof of such an origin for the disc is found in a number of stages like those figured, showing the

gradual concentration of the nuclei from all sides toward the embryonic area. The disc is early outlined as a result of this process, especially along its posterior border; but for some time, after this it occupies a larger area than is the case when it is completely formed. This is significant since, if it were merely a question of cell multiplication, the boundaries of the disc would be expected to gradually extend out from the centre, and not contract toward it. The nuclei of the disc in all of these stages are of the same size as those of the rest of the blastoderm; and this is true in later stages, as in fig. 15, where the disc is formed and the amnion has made its appearance. High powers show no distinction between the cells of

thin disc and those of the blastoderm outside of its circumference —

This fact evidently supports the view here advanced that the disc is formed by a concentration of the blastoderm cells — a rapid multiplication of cells within a restricted area should, it seems, soon lead to a marked difference in size between the nuclei within that area and outside of it — Dividing nuclei are found all over the ventral surface of the posterior end of the egg, during the stages leading to the formation of the disc; but they are not more numerous in the region finally occupied by the disc, until a comparatively late stage in its development represented by figs. (18) and (19) — At so late a stage more dividing nuclei would be expected in this area, since the cells

have become rather closely crowded — It must be remembered though, that the area occupied by the disc at this stage is still further reduced later — (see figures) The first rudiment of the embryo is certainly not formed from a number of discrete centres, as in the seapods and in certain insects — The concentration leading to its formation is, from the first, most marked in the posterior border of the disc, which early becomes sharply defined, as the cells draw together in concentric rows from the dorsal surface — The lateral edges are next involved, but much later, when the disc is otherwise well outlined, and its cells are quite closely crowded, the cells of the anterior end are not yet drawn together — Figs (34) and (35) — When the amnion is about to close over, the cells

of the end have drawn together and become incorporated in the disc —

I can not determine whether the concentration is accomplished by the migration of independent amnestrial cells toward the embryonic area, or whether the blastoderm is a continuous membrane of cells which contracts toward that point — The latter may be the case, though it is not clear that there is protoplasmic continuity, except where the cells are crowded in the region of the disc — It is however often difficult to establish continuity much later, where there can be no doubt of its existence in the serosa —

A less marked concentration of the surface cells has been observed in other insects, (notably by Patten (19) fig. 5. Pl. VIII and by Wheeler

(23) figs 63 and 64, and 66, and 68) resulting in a closer approximation of the cells of the embryonic area — In Ito Semiti, where the embryo is a comparatively small disc in its first appearance, the concentration to establish this disc is an especially marked process — McMurrich has discovered a similar method of the formation of the embryonic rudiment in Isopods — His figures (16) fig. 17, (18) & 19) and (50—52) show how the germ-band is formed in these crustacea, by a concentration of the surface cells toward the ventral side of the egg — He finds an intimate connection between this phenomenon and the formation of an inner-layer, and my observations on the Semiti's egg lead me to a somewhat similar

conclusive for it — Hence the detail with which I have described the process —

Sections show that the disc, in its early stages, is typically one layer of cells with a few scattered yolk-cells beneath — as the disc becomes more distinct however considerable displacement is observed in the nuclei of this layer, (it should be noted that this takes place at all points in the disc, laterally as well as near the middle line —) due to the concentration of the cells toward the centre — Hence some cells are crowded beneath the surface, as is shown in fig.(23), which is a true section of a very young disc (fig 16) — The crowding is of course more marked in cases of specimens where, as shown from the surface fig.(22), concentration has been carried to an extreme —

Here the disc is several layers thick in section, and its nuclei exhibit all stages in the branching beneath the surface — Fig. 24, — Sections of the embryonic area in which the disc is first distinctly outlined, show also that the nuclei are dividing in places to separate cells from the lower surface — Fig. 23) — These cells adhere in the lower layer of the disc —

<u>Under-layer and Embryonic Membranes.</u>

Figures (25¹), (26¹), (26²), (27) and (28) throw light on the processes now to be described. — Around the centre of the disc, the cells are now more closely crowded than elsewhere on its surface, and form a conspicuous dark spot — The figures also exhibit another circ— This is, the first trace of the amnio-serosa

fold, which appears from the surface as a
dark semicircle of crowded nuclei, arranged
concentrically along the posterior edge of the
disc — The rest of the disc looks much
lighter from the surface, the nuclei being
further apart — The figures represent
several stages in the growth of the central
dark spot, and in the increasing distinctness
of the posterior semicircle of nuclei —
Sections through these stages, and those just
preceding and immediately following,
taken in connection with what has been
learned from surface views, give interesting
data as to the formation of the mesoderm layer
and the amnion —
Mr Wheeler found, in working with Blattidae,
sections through these stages are very difficult
to obtain — It is impossible to get good
views of the entire egg, without adopting the

tedious, and not altogether satisfactory method of painting each section before cutting as Heider suggests (¹) — The method which I have used most is that first described by Dr Wm Patten — It is rather a slow process, but most satisfactory — In fact it can not be dispensed with, in handling such small objects; and Dr Patten has certainly done a great service in publishing it — In applying this method to sectioning the early stages of the Termite embryo, I break off most of the yolk in clove oil with sharp needles, under the dissecting microscope, and stick the portion to be sectioned in a piece of ruled paper — after imbedding in paraffin, the paper may be stripped off, leaving the object just only oriented for cutting — It is possible, in the case,

to obtain sections of quite early stage, in any plane and of any desired thinness — enough yolk adheres to the disc to make plain its relations to the interior of the egg —

I have studied the origin of the under-layer with special care, on account of the recent conflicting results of Wheeler[29] and Reymond[20] in regard to its formation in the Osteoptera — In the Termite there is no particular invagination — The under-layer begins to make its appearance at an early stage in the formation of the disc (figs. 16 and 19), when its cells are not very closely crowded — At this period, at irregular points in the embryonic area, lateral as well as median, some of the cells, as has been said, are pushed below the surface by the concentration of the

Blastoderm — Other cells are separated from the under surface of the ectoderm, at this period, by tangential divisions of its nuclei here and there —

As these processes continue, the under-layer constantly gains in importance — Its formation seems due chiefly to the concentration of the disc, and when this has reached the stage represented by fig.(27) the under-layer cells have, for the most part, collected into a plug projecting into the yolk — From the surface, this plug appears as a darkened area of crowded nuclei — A sagittal section of a disc of this age, shows the plug quite distinctly — It also is seen in this section that the under-layer cells are not confined to the plug, but are being pushed below from the ectoderm posterior to the plug, as far back as the hinder border of the disc fig.(28) A cross

section of the blastoderm through the
region of the sinus, at this stage, fig.
(30), when compared with figs. (31)
and (32) cutting the same region of
younger discs, ((31) stage represented by fig.
() and (32) of that shown in fig.)
shows that the plug has grown considerably
by the gradual additions of cells from
the ectoderm — Both surface views
and sections of these stages agree in
exhibiting no gastrula invagination — On the
contrary, it is as I have stated: the
under-layer arises at all points in the
germinal disc, as a result of a concentration of the embryonic area and of the
tangential divisions of its cells — The
formation of a mesentoblastic plug is
apparently a further outcome of the
concentration — (See Dr Minick on the formation

of the nuclear-layer in Isopods (1/2))—
Preparations of a series of discs, after the
nuclear-layer has mostly become crowded
into a plug, illustrate the growth of
this collection of cells — Figures (20), (27), and
(33–35) exhibit, in surface views, the
gradual extension of the plug, up to the
time when the amnioserosal fold has
grown well forward over the disc —

I have devoted much attention to the early
history of the embryonic membranes, on
account of the general interest which their
presence excites. — When the amnio-
-serosal fold is first clearly defined, as a
fold, in section, it appears from the
surface as is shown in fig. (34), as a semilunar
fold along the posterior border of the
embryonic disc, extending forward to

near the anterior end, on either side — Sagittal sections of this stage make it plain that the inner or amniotic layer of the fold is not to be distinguished from the ectoderm of the germinal disc [fig. 35]. It is of the same thickness as the ectoderm, and its nuclei are arranged in the same way (inverted of course) — The outer or serosal layer of the fold, on the other hand, is quite different [fig. 36]. It is a thin membrane of much flattened cells, with nuclei comparatively far apart — This is however not the first stage in the formation of the amnio-serosal fold of the Iguanite — Several stages before a fold can be made out in sections, its position is outlined on the surface of the disc — When the much-layer plug first appears in surface view, the embryonic disc is quite sharply marked out, covering

on its posterior border (fig. 25) — As has been said, it is along this border that the annuli do to appear — At first the nuclei in the edge, which have the same character as those of the rest of the disc or of the extra-embryonal blastoderm, give no indication of the fold about to arise there — In a slightly older disc, fig. 27, however, the posterior border is seen to be formed of a semicircle of closely crowded concentric rows of nuclei, which stands out very distinctly in surface views — Figure (28) represents a later stage than this and one before a fold can be distinguished in sections —

Sagittal sections of discs in these stages, up to fig. (36) teach that the posterior margin, corresponding to the dark

semicircle on the surface, differs from
the rest of the disc only in a somewhat
greater thickness. fig (37).—
It is the posterior, thickened margin, 2.
below, which lies folded over in fig (36)
and becomes the amnion.—
It will be noted then, in reference to the
origin of the amnion, that it is a portion
of the embryonic disc, and that it is
indicated as a thickening of the posterior
margin of the disc, before it folds over
to become the amnion.—
This agrees essentially with the figures which
Bruce gives of Manis (6) figs (42) & (43) of
plate II, with Patten's description and figures
of the Phrynosoma (19), with Will's of
the Aspidites (25), and with the results of
most other authors; though all are not
agreed to regard the amnion as a por-

of the embryo. Be this as it may it is certainly a portion of the first rudiment of the embryo of the Limulus. I have reserved a final section of this paper for a consideration of the origin of the membranes.

Preparations of eggs illustrating successive stages in the closing of the amniotic cavity, show that this is accomplished by the single, semilunar tail-fold from the posterior end of the disc. There are no separate lateral folds, nor is there any "head-fold". In a series of specimens the membranes are found extending further and further forward over the disc. Finally the amniotic cavity remains open at only a single point, at the anterior extremity of the disc. The closure of

which opening completes the process —
Figs.(35),(38), 39), (40),(40ª), and (41) —
During and just after the closure of the membrane, the amnion is not distinguishable from the ectoderm of the germinal disc — It is quite thick being composed of cells arranged into two layers, and its cells look like and act like those of the ectoderm — When the amniosomal fold has once arisen, the enclosure of the disc by it, appears to be due simply to the multiplication of the cells of the amniotic layer —

A sagittal section of an embryo just after the closure of the amniotic cavity shows this similarity strikingly fig.(42) — Comparing fig.(41) with figures of earlier stages, it is evident that considerable changes have taken place in the shape

of the embryo — The disc has now grown larger — It is about twice as long as broad and, while the posterior end is enlarged and rounded, the anterior extremity is distinctly pointed — The cells of this disc, and of the amnion, have become much smaller from repeated divisions, while those of the serosa have long since ceased to divide and are now very large — The sagittal section just referred to, fig.(y 2), brings out well the condition of the lower layer — The plug of cells which forms this layer has now become very prominent and lies beneath the anterior portion of the disc — It does not extend quite to the anterior extremity of the embryo and falls somewhat short of the posterior end —

The growth of the germ-band, its segmentation,
the appearance of appendages
and
— Revolution —

A series of embryos from the stage just
described to one exhibiting the first traces
of segmentation throw some light on
the method of growth — This is accomplished
by the hinder end of the embryonic band
growing back over the posterior end of
the yolk mass, just beneath the serosa,
while the head end remains fixed — Until
it reaches about half way over the
posterior pole, no marked change takes place
superficially, except an increase in
length and breadth — Figs. ( ) ( ) The
anterior end however becomes gradually
less pointed — When the posterior

end of the band have pushed around the yolk-mass, and up about one third of the flattened dorsal side of the egg, the embryo forms a U shaped clasp over the enlarged end of the yolk — Fig. 45 — At this time the band is still unsegmented — Posteriorly it terminates as before in a rounded extremity — The anterior end, however, has gradually undergone considerable change — From being a narrow pointed tip to the band (fig. 41), it first widened into a square end, figs. (43) and (44), and finally spread out over the yolk anteriorly and laterally (fig (45), until now, fig. 46, this region has become the most prominent part of the embryo — Anteriorly, just in front of the place where the mouth is to appear the cephalic region is slightly emarginate — On either side it extends up on to yolk, in a broad lobe with rounded borders. fig...

In sections of these stages several facts become
apparent — The under layer is gradually marked
off from the ectoderm — As the ectodermic layer
of the embryonic band elongates, the under-layer
cells multiply and follow this growth posteriorly
fig. 67 — The folds which in earlier stages marked
the anterior end of the embryo, now ends
in front at the base of the cephalic region,
fig.(67, 68) — The cephalic region is thus seen
to have been formed, entirely by the spreading
of the ectoderm anterior to the under-
-layer (?), after the closure of the embryonic
membranes — As the germ-band has elongated,
the under-layer has grown steadily back,
apparently largely as a result of the
multiplication and rearrangement of its
own cells — fig.(62) and (67) — The ectoderm grows
somewhat more rapidly than the lower-layer
which does not extend quite to the posterior

end of the embryo during its elongation. When the posterior end of the band has grown about half-way over the end of the egg, the "mesentoderm" is collected for the most part under the tail end of the band — Fig (8) — it forms only a comparatively thin layer anteriorly where it is quite distinct from the ectoderm. At the other end of the embryo, however, I am not certain, as yet, that no cells wander from the ectoderm below, to add to the under-layer — it is possible, I think, that such is the case — I have not followed the differentiation of this layer further, but may state that the entoderm is not found as a distinct layer until after the appearance of segments. An origin of the entoderm from yolk-cells is not possible, since there are only very few of

than in the yolk and much earlier they soon become larger than the entoderm cells ever are — Fig. 8.

These sections also exhibit well the changes which take place in the growth of the amnion — up to the time of the closure of the membrane the amnion was seen to be strikingly like the ectoderm of the disc — now, as the germ-band elongates posteriorly and in the cephalic region pushes out anteriorly, though the cells of the amnion continue to multiply, the membrane is gradually stretched, the cells arranging themselves in a single layer — This takes place first at the anterior end, the posterior end for sometime retaining its early appearance — Figs (42), (57) and (58) —

The first traces of segmentation and appendages appear suddenly, at about the last stage described, where the germ-band has become a U shaped cap over the posterior end of the yolk mass. The earliest segmented band I have yet found, is one in which the antennae are just noticeable as backward processes of the cephalic lobe post-oral in position — Figs. (79) and (79a). The first maxillary and first thoracic segments are seen. The other anterior segments are indistinct and the tail-piece is long and unsegmented. — The anterior segments, through the first thoracic, appear then to have arisen almost simultaneously. — There are no macro-somites as Graber described for Stenobothrus and other forms. (8) and (?) — Later

embryos show a progressive increase in length and complexity of the germ-band. When the band has pushed forward along the surface of the yolk, almost to the anterior end of the egg, three additional segments have been added. Fig. 50. These are the two posterior thoracic segments and the first abdominal. They are added successively from before back, as I have embryos in which the first thoracic is the last segment distinguishable, others with an indistinct second thoracic besides, and yet a third lot with three distinct thoracic segments and an indistinct abdominal. In older trunks more abdominal segments are successively added posteriorly. The deep inter--segmental groove separating the first

and second thoracic segment is a peculiarity of this stage — Later, when the appendages become marked this characteristic disappears —

The head of such an embryo is shown from the dorsal surface in figure (51) in an early stage of folding off from the yolk — It will be remembered that, in the first appearance of segmentation, the cephalic region was formed of two broad lobes, which had arisen as expansions of the anterior end of the germ-disc — Soon, the lateral margins of these lobes bend (or rather roll) up dorsally and fold over toward the middle line — This is the beginning of the formation of a head-cavity — In fig (51) the cephalic lobe is seen to have formed a sort of pocket between its dorsal and

ventral folds — In older embryos this process is continued, the lateral folds of the head region finally meeting over the dorsal mid-line and enclosing the head-cavity — The antennary rudiments are included in this folding; and hence, later, their cavities are continuous with that of the head — At an early stage in the formation of the head, the labrum appears; and, as far as I can determine at present, it is from the first an unpaired, median outpushing, just over the mouth opening — This median fold appears to assist in drawing the head together into its definitive compact size and shape —

The embryo has now attained a stage like that figured by Brandt for Calopteryx in fig. (11) of his paper (3) — It is

more like Corixa however, since unlike the embryo of the Libellulid it is not surrounded by the yolk — (I must refer here to Graber's figures of the development of Stenobothrus,(?) which is remarkably similar to that of the Termite, as described here in all of the general features, though no "revolution" has been described for stenobothrus.) In the Termite when the germ-band has grown along the surface of the yolk to the anterior end of the egg, the posterior portion of the abdominal region sinks slightly into the yolk — As the embryo continues its elongation, this bend in the abdominal region becomes more marked, and the tail-end of the band coils ventrally toward the head until the stage represented in fig (52) is reached — The appendages have become prominent —

The 1st maxillae are tuberled, and the 2nd pair less markedly so — In the abdominal region 10 well marked segments have appeared and each bears a distinct pair of appendages — (No abdominal appendages are figured by Brandt at the corresponding stage of Calopteryx — none of Grabers figures of Stenobothrus exhibit such marked rudimentary appendages in the abdominal region —)—

From this stage until "revolution" the embryo undergoes no change of shape or position; but the sides of the body grow dorsally somewhat, and the appendages elongate considerably —

Revolution is accomplished as described by Brandt (3) for the Libellulid; and when it is completed, the head of the embryo has shifted up

to the anterior end of the egg, while the tail-end lies beneath the micropyle — The ventral surface of the embryo is now on the micropylar side of the egg, as at first up to the closure of the embryonic membranes —

I have not, as yet, studied in detail the stages after the appearance of the appendages; but will state, in regard to the origin of the central nervous system, that is evident from sections that it arises from neuroblast cells, as Wheeler(²) and Viallanes (²⁶) have described for the Orthoptera — In the Termite this process is very marked — I may say, with respect to the reproductive organs, that there is no trace of them distinguishable in the embryo within

the egg, or just after leaving the egg membranes — Fritz Müller (18) found that it was impossible to detect any trace of sexual organs in the soldiers, and workers of certain species of Termites. This is true of the species from Jamaica on which I have worked, (Eutermes - *(Rippertii ?)) — a careful study of sections of both the larvae and adults of this form fails to reveal any trace of sexual organs — I am now investigating this question further —

* The Jamaica form in some respects differs very much from Rippertii, hence I have placed an interrogation mark after the specific name.

— The Mesentoderm —

Recently the origin of the under-layer in what are regarded as the most primitive insects, the Orthoptera, have been carefully studied by two well known investigators, who have reached quite contradictory results. Wheeler,[21] in his "Contribution to insect embryology" has devoted considerable space to a review of this question. His conclusion is expressed as follows: "It follows from the observations here recorded, fragmentary as they are in many respects, together with Graber's observations on Stenobothrus, that the Orthoptera can no longer be regarded as hors de ligne, so far as the formation of their germ-layers is concerned. In all the families of the order, save the Phasmidae, an invaginate gastrula is here found, and there can be little doubt that the investigator

who is so fortunate as to study embryos of
the family will find in them essentially
the cause" process of germ-layer formation—
The view is now pretty generally held that
in the insecta both mesoderm and endoderm
arise from a median longitudinal furrow,
(the inner layer throughout nearly the entire
length, the latter only in the oral and
anal regions of the germ-band) and
that the vitellophags, or cells left in the
yolk at a time when the remaining cleavage
products are travelling to the surface to
form the blastoderm, take no part whatsoever
in the formation of the mesenteron, but
degenerate in situ and finally undergo
dissolution."
I have been unable to obtain a copy of Heymons's
study on the germ-layer formation of the Hemiptera
and Dermaptera,[13] but in conclusions he

appeared in abstracts and are as follows:
The yolk-cells take no part in the formation of the embryo — There is no true gastrulation process, the middle-layer arises from all parts of the embryonic area — When a typical gastrula invagination occurs, as in most insects, it is to be explained as a simple mechanical process caused by an aggregation of cells at one point. The layer generally known as the mesentoderm is in reality only mesoderm, the entoderm appearing relatively late and arising from the ectoderm of the stomodeal and proctodeal invaginations —

My observations plainly accord rather with the latter view than with Wheeler's, which is the one more generally accepted — The Termite, which is certainly as primitive as any other form hitherto described, exhibits no gastrula

invagination — I have shown that the under-layer
begins to appear at all, even in the embryonic
rudiment, in an early stage in its formation —
The plug of lower-layer cells, which becomes so
prominent as the germinal-disc grows more
distinct, is apparently the outcome of secondary
conditions — The relation of such a method
of formation of the under-layer to that generally
described for insects is interesting to
consider — This process does not appear to me
to be derived from an invagination, as
a closed gastrula — It is rather a method
of delamination, where there is a further
tendency for the lower-layer cells to collect
toward the middle line — A similar
method has been described in arachnids
and myriopods, and all of these facts
taken together give some weight to
Heymons's view — I should not however

agree with him in regarding the invaginate gastrula as simply a mechanical process, but should prefer to look on it as arising by natural selection, from the method of determination just described —

Whether does not claim to have made a special study of this question but, since his opinion agrees with what has generally been thought to be true, further comparative research would seem necessary, before we finally decide that the so called invaginate gastrulae of insects are secondary phenomena — I am not inclined to regard the question settled by what I have found in the Termite alone, but think that the Libellulidae, and other primitive forms, should be again studied from this point of view —

As to the inclusion of the Termite I must say that it appears late, after the

germ-band is segmented. The yolk-cells can take no part in the formation of this layer, since at an early stage, before the closure of the amniotic cavity, they have become very large and unlike the cells which later form the endoderm.

-- The amnion --

The origin of the amnion, whose appearance in the insect embryo is perhaps the most striking peculiarity of the development, has been much discussed. At the present time however the question is still quite unsettled, opinions being divided chiefly between two views.
Wheeler[(21)] supports a purely mechanical theory and holds that "the amnion first made its appearance in the ancestral Thysanura." Korschelt and Heider[15], on the contrary, have

voted considerable space in their text-book to an account which refers this occurrence to a phenomenon observed in myriopods. The view, which has received more attention than that of Ryder and Wheeler[20][21] maintains that there are two main types of development among insects — "The imaginate" type is best seen in the Libellulidae, which "represent the direct connecting link (anschluss) with the phenomena exhibited by the myriopods, and hence must be regarded as the more primitive type — Judging from Heider's remarks in his monograph on Hydrophilus[12] no genetic relationship is meant here between the doubling-up of the myriopod embryo and the origin of the amnion of the insects; but the treatment in the text-book certainly suggests strongly such an interpretation —

It seems to me, according to the expert that this is true — Diagnosis from [Zuwachs?] figures of the development of [Selbstenys?] emphasize the conformance with the [hypothesis?], which is insisted on in the following terms — "Wir haben oben gesehen dass bei den [typischen?] [fortschreitenden?] Längenwachsthum der Keimstreif derselbe in seiner Mitte eingeknickt und in das Innere des Eies versenkt wird. In dieser Einsenkung, welche wir uns zunächst durch das räumliche Missverhältniss zwischen dem langgestreckten Keimstreif und der rundlichen Eiform entstanden zu denken haben, werden wir [wie dies schon Grube andeutete und Will neuerdings aus-führlicher begründet hat] den Ausgangspunkt für die Entwicklung der [umgänzten?] Keimstreife der Libelluliden

zu suchen haben — Wir werden demnach
für die Entwicklung der InsectenKeimstreifs
die Form der Invagination als die
ursprüngliche betrachten —

This account is apparently based on
Heider's[12] discussion of the subject in
his monograph on Hydrophilus — It is
a modification of Will's[175] theory, against
which in its original form, Graber,
in a more recent paper[9] than the one
referred to, brought forward strong
objections —

Since the publication of Korschelt and Heider's
text-book further investigation has shown
that a number of the Orthoptera besides
Oecanthus, as well as the Termite, exhibit
essentially the same phenomena of
development as the Libellulidæ — It now

seems evident, that there are no grounds
whatever for regarding the method
of development followed by the Libellulid
embryo as one which more primitive than
that of Oecanthus, Xylleta or the Termite —
There is some reason for believing it to
be more secondary than any of them,
since the embryo be of the 'immersed
type — A superficial germ-band is
generally characteristic of Arthropods, and
where we find one sunken into the
yolk, there is cause to believe the position
has been assumed secondarily —
Among the insects, most forms, [and
especially the Orthoptera and Termites
which are probably equiphrenally primitive,]
agree in having superficial embryos —
The exceptions are rather marked and
are found among the Lepidoptera, the

Hemiptera and the Libellulidæ — In the Lepidoptera the immersed position is admitted to be secondarily derived from the superficial for protection, nutrition, or some other cause — It appears to me most probable that the same is true for the Libellulidæ and Hemiptera with inner germ-bands — Hence I should regard the superficial embryos of the Orthoptera and the Termites as more typically primitive for insects, and should expect to find among them the best examples of the ancestral method of forming the amnion — Similar reasons may have brought about the "sinking-in" of the myriopod embryo into the yolk and the origin of the membranes in insects, but the Libellulidæ do not appear to represent the latter process

in its most primitive condition — It seems probable that the amnion arose first to cover a superficial germ-band, and that the Libellulid and Lepidopteran embryos have later taken up an "immersed" position, while now insects have retained the primitive superficial one —

As to the cause assigned for the sinking of the Myriopod band into the yolk, the resistance offered by the spherical chorion & the growth of the elongating germ-band, expressed as "das räumliche Missverhältniss zwischen dem langgestreckten Keimstreif und der rundlichen Eiform", it may be observed that similar conditions in the arachnid egg and in that of a fish do not lead to such an invagination — (see also Graber's criterion of the idea

in his monograph on the germ-bands of
insects &c.)—

A striking character of the development of the
Termite is the small size of the first
rudiment of the embryo, the germ-disc, when
compared with the definitive length of
the embryonic band — The primary rudiment
much elongates through the whole length of
the egg, and add successively all its
segments of the body, before the embryo is
formed — This is equally noticeable in
the case of some of the Orthoptera; but is
less marked in most insects, especially
among the more specialized forms of the
group —

             Now it seems to me that
the Termite and the Orthoptera with a
superficial embryo- beginning in a disc,

which must elongate considerably to attain the definitive number of segments, can adhered most nearly to the typical method of development for Arthropods, and probably best represent the development of the ancestral insects — The fact of the development of the Crustacea, the Palaeostraca, the Arachnida, and to a less marked degree the Myriapods show a similar disproportion in size between the primary rudiment and the definitive segmented embryo — This may be illustrated by referring to the growth of a nauplius to its adult form —

These primitive forms of insects are also characterized by another peculiarity of interest in the present discussion — The amnion arises very early and completely covers the embryonic disc soon after its appearance — We do not know with certainty to what need of the embryo

the amnion responds, but would surely exist
to find it in its most primitive condition
in the very forms under consideration — I
believe that is the case and that insects, in which
the membranes become prominent and cover over the
embryo comparatively late in its growth represent a
secondary condition — If, as is generally supposed,
the amnion serve as a protection for the germ-band
against mechanical injury or too rapid evaporation,
it would have been a great advantage for it to
appear at the earliest possible moment in
the growth of the embryo — The moment is when
the germ-disc is established and about to grow into
the elongated, segmented embryo — From this time a
superficial germ-band would be constantly exposed
to the dangers mentioned — Hence it seems
reasonable to suppose, that the embryos
of the ancestral amniote insects became
covered by the membranes at

this early stage in the development — Among primitive forms, the Termites and some of the Orthoptera (Stenobothrus, Gryllus &c) have best retained this method of the formation of the amnion — In other Orthoptera, the Libellulidæ, some of the Hemiptera, and many other insects the ancestral history is not so well preserved — In these the amnion no longer closes over at such an early stage — Wheeler's figures of the germ-bands of Blatta and Doryphora (23), Graber's of Lina (9), Heider's of Hydrophilus (12), and Weismann's of Chironomus illustrate this point for superficial embryos — The Libellulidæ, and some of the Hemiptera, show to a marked degree ancestral characters. However, the much retarded closure of the amniotic cavity, and the presence of the so called secondary hind-field, together

with the secondary nummered position of the
gene-band, render these forms the
typical examples of the probable primitive
method of development —

Wheeler has recently (?¹) adapted Ryder's (²)
theory of a mechanical origin for the
annuism of vertebrates to the insect annuism —
Of course the word mechanical is used here
in its narrower sense, refining the question
to a few simply stated conditions of
genuine and mechanical strain — This view
must be seriously considered, it seems to me
only because of the habitual caution which
characterizes Wheeler's work, and the more
tentatively nature of his conclusions —
The theory is clearly stated in the following
quotation — — — — "The nummeral field is
a mechanical result of a local multiplication

of the blastoderm due to rapid proliferation in a single layer of cells." Here is the vesicular one-layered blastoderm filled with yolk, and the germ-band arising by rapid proliferation at one point. The resistance of the yolk being less than the external resistance of the tightly fitting chorion and vitelline membrane on the one hand, combined with the peripheral resistance of the extra-embryonal blastoderm on the other, the germ-band is forced to invaginate. The invagination is favored by the displacement of yolk during its liquefaction and absorption by the growing embryo. We may suppose that this invagination which results in the formation of the amnio-serosal fold, assumed a definite and specific character in different groups of

insects—"

It would be an extremely nice problem to determine, with even an approximation to accuracy, the mechanical forces at work in an insect's egg where the amnioserosal fold arises, and still more difficult to be able to predict what must happen as a result of their action— The above quotations only make this evident, and can not, in any sense, it seems to me, be accepted as an adequate explanation of the origin of the amnion of insects—

It seems sufficient to recall the fact that there is no amnion among the Crustacea, nor among the Myriopods or Apterygota, and that it is lacking in certain of the higher insects— if such

The same conditions of pressure are brought to bear on the embryonic area of the Crustacea, Arguispoda, or Apterygota as are claimed to force the invagination which forms the membranes of the Pterygota — But no amnion is formed in any of these forms — It will hardly be claimed that the invaginations which occur in the rapidly proliferating areas of the teleost blasto-derm are the direct result of mechanical strains — This indicates other factors in their origin, which I believe are also present in the case before us —

In those highly specialized insects that entirely lack an amnion, its failure to appear is even more marked — Here, within the same group there are forms which, in the face of the forces above stated as efficient to

produce an amnion, form none —
I may say here that my own observations on
the Termite do not support Wheeler's view.
The germ disc is in all probability, as I have
tried to show, not only the result of a
rapid proliferation, but also of a concentration
or wandering of cells toward the point where
it is to arise — The same tendency has been
shown to be present in the formation of the
first rudiment of the embryo in other
insects, and in other arthropods — Is this
phenomenon in no way connected with
the origin of the amniocaecal fold? It
is a factor of a very different nature from
those relied on by Wheeler, but may be
equally if not more important — If such
is the case, the origin of the membranes
can not be so directly referred to
simple conditions of pressure —

I should say then, that, though Wheeler may have enumerated some of the most apparent mechanical forces present in the egg of an insect he omits to mention a more important factor in the origin of organic structures — This is adaptiveness — It may be admitted that in the ancestors of amniotic insects the physical constitution of the egg was favorable to the origin of an amnion, but before it could arise a further condition seems imperative, a necessity for such a structure must be felt — Such a problem may with great probability be justly spoken of as a mechanical one in its ultimate sense — We have become accustomed to regard all natural phenomena as having, in all probability, fundamentally a physico-chemical explanation — But granting this, when an attempt is to

is made to explain the origin of a structure, account must be taken of all the conditions. Besides the mechanical forces mentioned by Ryder and others there are apparently, as has been said, other factors without which no ammion would appear — The additional influences leading to the origin of an ammion in the winged insects, must there be the conjunction of certain physico-chemical forces not brought into play in the case of the apterygota (for instance), which are too subtle to be defined except in most general terms — we can only indicate their presence by saying that the nascent structure must be a response to a lack of adjustment of the organism to its environment — The mechanical nature of the problem can not, I think, be stated in a more

definite manner than as follows: For some reason (perhaps for protection against injury as Ruschell and Heider suggest) it became a necessity that the superficial germ-band of the ancestral pterygota should be covered over soon after it appeared — The conditions in the egg being favorable for the formation of such a membrane, when the environmental influences became sufficiently modified an amnion was formed, by a folding of a portion of the small disc beneath the serosa, as an adaptation to these influences — In those in which this did not occur were of course eliminated — Where found some among the higher insects, an adaptation to special conditions. The early completion of this process became

are important, and in a few cases this lead to the degeneration and disappearance of the aurium —

N.B. The figures to illustrate this Thesis are submitted separately, as I am obliged to keep them for printing —

Special reference must be made to the following authors:

1. Ayers. H. On the development of Oecanthus niveus and its parasite Teleas.
    Mem. Boston Soc. Nath. Hist. vol 3. 854 —

2. Bobretzky. N. Ueber die Bildung des Blastoderms und der Keimblätter bei Insecten.
    Zeitschr. f. wiss. zool 31. Bd. 1878 —

3. Brandt. A. Beiträge zur Entwicklungsgeschichte der Libellulden und Hemipteren.
    Mem. acad. St. Petersburg (7) Tom 13. 1869 —

4. Brauer. Fr. Systematisch-zoologische Studien
    Sitz. Isl. akad. wiss. Wien 2. Bd. 1882

5. Cramer. C. Beiträge zur Entwicklungsgeschichte der
    Herpinia —
        Zeitschr. f. wiss. Zool. 32 57. 1893, 32 571

6. Brace L. T. Observations on the embryology of
    insects and Arachnids —
        a memorial volume — Baltimore 1887 —

7. Graber. V. Die Insecten — München. 1877 —
        ds.: Die Naturkräfte — 1879 —

8. Graber. V. ueber die primäre Segmentierung des
    Keimstreifs der Insecten —
        Morph. Jahrb. — 14. 1888 —

9. Graber. V. Vergleichende Studien am Keimstreif der
    Insecten —
        Denkschr. Acad. Wiss. Wien 57 Bd. 1890.

10 - Erlanger. V Vergleichende Studien über die Keimhüllen
und die Rückenbildung der Insekten.
Denkschr. k.k. Akad. Wien. Bd XX 1892

11 - Hagen. H. Page in Schfol. vol XX. see i. N.J.
Proc. Bost. nath. hist. soc -

12 - Heider. K. Die Embryonalentwicklung von
Hydrophilus piceus u. Jena 1889

13 - Heymon. R. On the development of Blattina
and Termopsis, with special reference
t germ. layer formation -
<u>Abstract</u> in Journal Roy. Mikr. soc. 1894

14 - Kishinouge. K. The development of Lumbris roy... bin.
mem... of Tokyo 1891

15. Korschelt und Heider.
   Lehrbuch der vergleichenden Entwicklungs-
   -geschichte der wirbellosen Thiere -
                    Jena. 1890 —

16. Mc Murrich - J.P.   Embryology of the Isopod Crustacea.
                    Journal of Morphology vol XI. 1895.

17. Metschnikoff. E. Untersuchungen über die
                Embryologie der Nemertinen -
                Zeitschr. f. wiss. Zool o Bd. 166 —

18. Müller Fr.   Beiträge zur Kenntnis der Amisten.
                Jen. Zeitschr. - Naturwissensch
                7 Bd. 1873 —

19. Patten. Wm. The development of Phyzoids into
    a preliminary note on the development
    of Blatta germanica.
            Quart. Journal. Micr. sc. vol 20. 18a.

20. Ryder. John. A.   The origin of the amnion.
            American Naturalist. vol.20. 886.
                pages. 79 - , 85 —

21. Wagner. J.   Formations in the formation of the
        germ layers, yolk cells, and adjoining
        membranes of Arthropods.
            Biol. Centralblatt. XIV. no 10. 85p —

22. Weismann. A.   Die Entwicklung der Cysten im
        Ei nach zerlechteyn im
        Chromoma of innere mutane und
        Belag Cuire.
            Zeitschr. f. Wiss. zool. 3.14. 105

23 – Wheeler. W.M. The embryology of Blatta germanica
and Xiphora decemlineata –
Journal of Morphol. vol 3. 189-

24 – Wheeler. W.M. A contribution to insect
embryology –
Journal of Morphol. vol 5. 173-

25 – Will. L. Entwicklungsgeschichte der viviparen
Aphiden –
Spengel: Zool. Jahrbücher.
Abth. f. Anat. und Ent. 3 Bd. – 1888 –

26 – Viallanes. H. Sur quelques points de l'histoire
du développement embryonnaire
de la Mante religieuse –
Rec. Biol. du Nord de la France
Tom. 2 – 1889- 1890-

Vita —

Henry McElderry Knower was born in Baltimore, Md. August 5th 1868 —

He received his elementary education at the late Mr George G. Carey's school — In 1887 he entered the Johns Hopkins University, electing the Chemical-Biological course, and received the degree of Bachelor of Arts in 1890 at that institution —

On entering as a candidate for the degree of Doctor of Philosophy, he selected Animal-Morphology as his principal subject, with Physiology and Botany as first and second subordinates respectively — While pursuing his graduate work he held the positions of assistant in Histology, assistant in Zoology, University scholar, and Fellow in Zoology, and during the past year was Adam T. Bruce Fellow in Zoology —

www.ingramcontent.com/pod-product-compliance
Lightning Source LLC
Chambersburg PA
CBHW020249170426
43202CB0C008B/285